This book is dedicated to those who inspired me to love fishing: Grandpa Archie Thompson, Cousin Stormy Joy, Uncle Timmy, Pa Willie Thompson, and all those who have fought and will fight for the salmon, the river, and keeping Native American culture alive.

Copyright © 2025 by Brook M. Thompson

All rights reserved. Except for brief passages quoted in a review, no portion of this work may be reproduced or transmitted in any form or by any means, electronic or mechanical, including photocopying and recording, or by any information storage or retrieval system, or be used in training generative artificial intelligence (AI) technologies or developing machine-learning language models, without permission in writing from Heyday.

The publisher gratefully acknowledges Humboldt Area Foundation's Native Cultures Fund for its generous support in the making and publication of this book.

Library of Congress Cataloging-in-Publication Data is available.

Cover Art: Anastasia Khmelevska
Cover Design: Brook M. Thompson and Marlon Rigel
Illustration Concepts: Brook M. Thompson
Interior Design/Typesetting: Brook M. Thompson and Marlon Rigel

Published by Heyday
P.O. Box 9145, Berkeley, California 94709
(510) 549-3564
heydaybooks.com

Printed in China

10 9 8 7 6 5 4 3 2 1

I Love Salmon and Lampreys

A Native Story of Resilience

Brook M. Thompson

Illustrations by
Anastasia Khmelevska

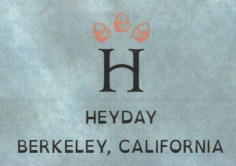

HEYDAY

BERKELEY, CALIFORNIA

I love fishing with my dad, uncles, and cousins. They show me our family's fishing spots and tell stories of their youth.

I cook salmon and lampreys with my family.

We make jerky in our traditional smokehouse. We bake salmon on redwood sticks. We make burger patties, stew, and barbecue too. The salmon feed us in the spring, summer, and fall. The lampreys feed us in the winter. Salmon and lampreys help my body become healthy and strong.

I admire how salmon and lampreys go from the river into the unknown ocean. They grow up in the ocean and then come back home to the same stream they were born in to have their kids.

Lampreys have been on Earth since even before the dinosaurs and sharks! I think that's cool. Also, I like that lampreys catch free rides on salmon by the suction cup and teeth in their mouths.

Today, these salmon are the great-great-great-great-great-great-grandkids of the salmon who had a relationship with my great-grandparents.

Dams on the river had made the water unhealthy for the salmon. I was so sad. The salmon are my friends. Salmon feed me, my family, and my friends. I want to protect them.

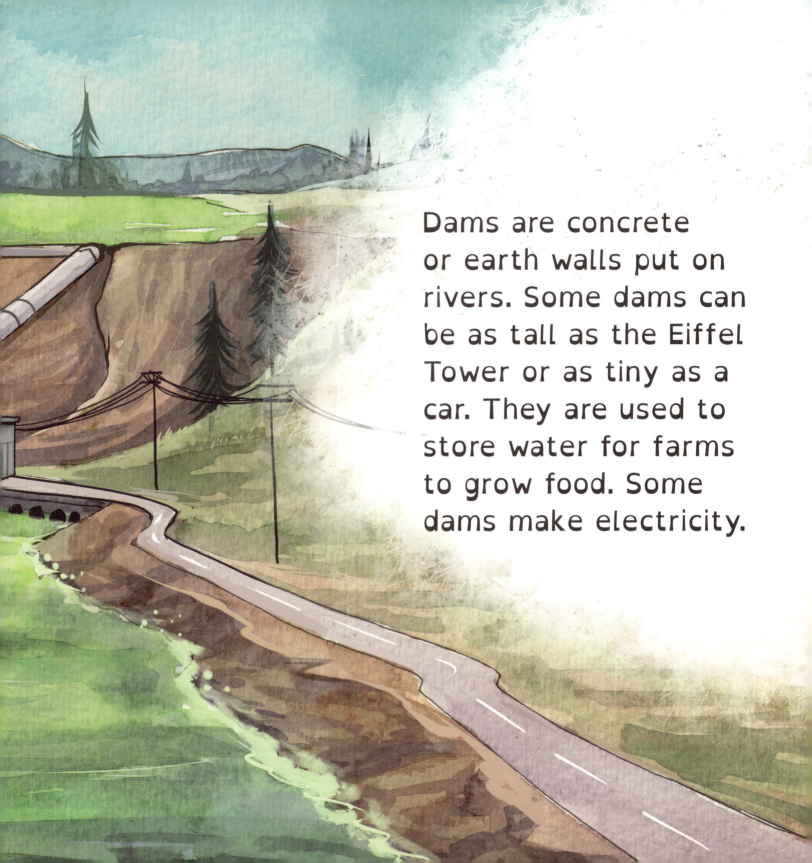

Dams are concrete or earth walls put on rivers. Some dams can be as tall as the Eiffel Tower or as tiny as a car. They are used to store water for farms to grow food. Some dams make electricity.

Dams also block the salmon's way home. Dams can make the water behind them too warm for fish. Toxic green algae can form. The water gets too gross to swim in.

The tribal people and our friends picked up signs and marched, scientists did research, and the politicians made a fuss so that the salmon kill would never happen again.

We took our sad and mad feelings from what happened to the salmon and turned them into change.

Finally, our voices were heard. It took over twenty years, but in the end, we made an agreement with the lawmakers and the companies that owned the dams, and four of the six dams were finally removed!

Learning from the endurance of the salmon and lampreys, I grew up and began to explore a new world away from my home. I went to college to learn about dams and about how to protect salmon in the future. I became an engineer and scientist. I do fancy math and science to understand water, nature, people, animals, and buildings.

Then like salmon and lampreys, I made my way home to the Klamath River in Northern California.

Now I take care of fish for people in the future, like my ancestors did for me in past years.

I love salmon and lampreys.

FUN FACTS

- Despite looking like an eel, lampreys are actually fish!

- Chinook salmon are also called king salmon.

- Lampreys have a third eye on top of their heads, which is a spot that can sense changes in light and dark. Many lizards, reptiles, and frogs have this too. The third eye looks like a dot on top of their heads.

- Lampreys as a species are over 450 million years old! The T. rex is less than 90 million years old.

- Salmon and lampreys are anadromous fish, meaning they spend part of their lives in the ocean, which has salt water, and part of their lives in fresh water. Not all fish can do this.

- There are five types of Pacific salmon. The ones in the Klamath River are Chinook and coho.

Holding a Salmon

Cooking Traditional Salmon on Sticks

Iron Gate Dam, March 2024

Iron Gate Dam, August 2024

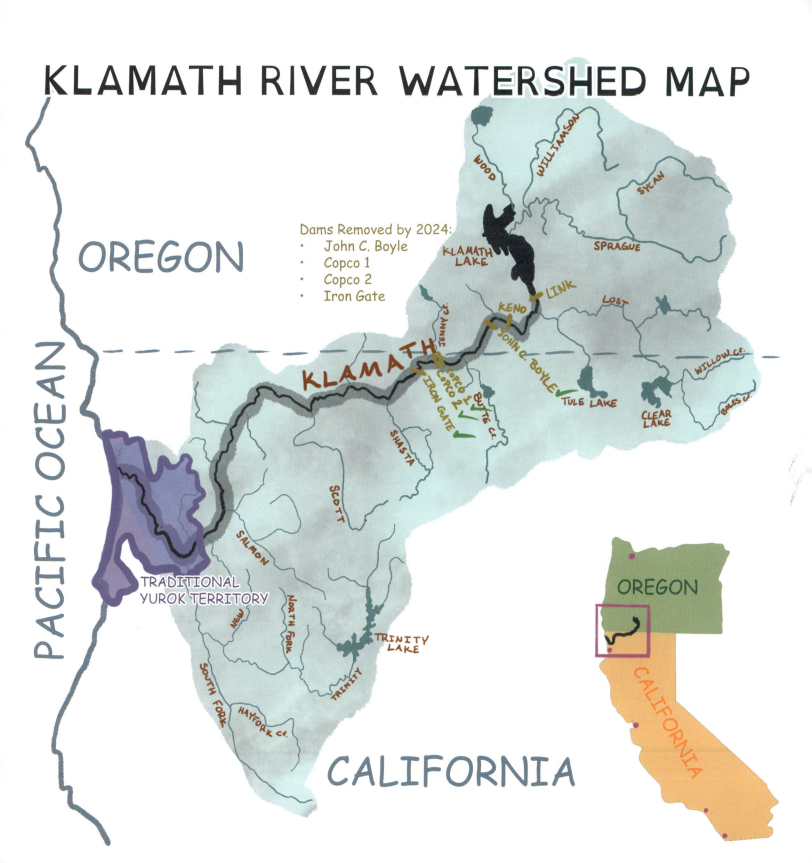

About This Book

I Love Salmon and Lampreys is based on the life of Brook M. Thompson. This book is about her experience growing up along the Klamath River in Northern California in the early 2000s when a deadly parasite infected the salmon. Brook's tribes have lived along the river for thousands of years, since time immemorial. Brook was inspired by the close cultural relationship that salmon and lampreys have to her tribes, but they have been in continued decline. After the removal of four dams on the Klamath River, she wanted to share this story of triumph in the face of an impossible task with all, so others can learn why we should protect the salmon and lampreys.

This book's text is written in a font called OpenDyslexic. The bottom of the letters are larger, which makes it easier for readers with dyslexia to read.

About the Author

Brook M. Thompson is a part of the Yurok and Karuk Tribes. She is a neurodivergent and Two-Spirit author with dyslexia. She has a BS in civil engineering from Portland State University and an MS in environmental engineering from Stanford University, and will soon have a PhD in environmental studies from University of California, Santa Cruz, where she studies water, politics, restoration, and salmon. You can find out more about her at brookmthompson.com.

About the Illustrator

Anastasia Khmelevska is an illustrator based in Lviv, Ukraine. She has illustrated several children's books, including *My Invisible Zoo*, *Marella the Mermaid*, and *Anything Helps*. Follow her on Instagram @cute_miuu.